SandCastle

What Should I Eat?

Proteins
Are Powerful

GRM°c1

Amanda Rondeau

Consulting Editor
Monica Marx, M.A./Reading Specialist

ABDO
Publishing Company

Published by SandCastle™, an imprint of ABDO Publishing Company, 4940 Viking Drive, Edina, Minnesota 55435.

Credits
Edited by: Pam Price
Curriculum Coordinator: Nancy Tuminelly
Cover and Interior Design and Production: Mighty Media
Photo Credits: Artville, Brand X Pictures, Comstock, Eyewire Images, Image 100, PhotoDisc

Library of Congress Cataloging-in-Publication Data

Rondeau, Amanda, 1974-
    Proteins are powerful / Amanda Rondeau.
      p. cm. -- (What should I eat?)
    Summary: A simple introduction to the protein food group and why proteins are important for us to eat.
    ISBN 1-57765-836-1
    1. Proteins in human nutrition--Juvenile literature. 2. Food--Protein content--Juvenile literature. [1. Proteins. 2. Nutrition.] I. Title.

TX553.P7 R66 2002
613.2'82--dc21

2002018365

SandCastle™ books are created by a professional team of educators, reading specialists, and content developers around five essential components that include phonemic awareness, phonics, vocabulary, text comprehension, and fluency. All books are written, reviewed, and leveled for guided reading, early intervention reading, and Accelerated Reader® programs and designed for use in shared, guided, and independent reading and writing activities to support a balanced approach to literacy instruction.

Let Us Know

After reading the book, SandCastle would like you to tell us your stories about reading. What is your favorite page? Was there something hard that you needed help with? Share the ups and downs of learning to read. We want to hear from you! To get posted on the ABDO Publishing Company Web site, send us email at:

sandcastle@abdopub.com

SandCastle Level: Transitional

What is the protein group?

Fats & Sweets

Eat LESS

MILK Group
2–3
servings

PROTEIN Group
2–3
servings

VEGETABLE Group
3–5
servings

FRUIT Group
2–4
servings

GRAIN Group **6–11** servings

*For suggested serving sizes, see page 22.

This is the food pyramid.

There are 6 food groups in the pyramid.

The food pyramid helps us know how to eat right.

Eating right helps us stay healthy.

The protein group is part of the food pyramid.

We should eat 2 to 3 servings from the protein group every day.

Protein is good for our bodies.

Protein helps us stay healthy.

There are many kinds of food in the protein group.

Protein helps our muscles grow bigger and stronger.

Protein gives us energy.

Did you know nuts are in the protein group?

Peanut butter is a protein because it is made from nuts.

A peanut butter sandwich is great for lunch.

Did you know beans are in the protein group?

Beans are vegetables that have protein.

That is why they are in both the vegetable and protein groups.

Beans are good in tacos or soup.

Did you know meat is in the protein group?

Turkey, chicken, fish, and hamburger have protein.

Turkey is good hot with dinner or cold on sandwiches.

Did you know eggs are in the protein group?

Most of the eggs we eat are from chickens.

But some people eat other kinds, like quail eggs!

Eggs are great for breakfast.

Can you think of other foods in the protein group?

What is your favorite food in the protein group?

Index

What Counts As a Serving?

Meat, Poultry, Fish, Dry Beans, Eggs, and Nuts	
2–3 ounces of cooked lean meat, poultry, or fish	½ cup of cooked dry beans or 1 egg counts as 1 ounce of lean meat. 2 tablespoons of peanut butter or ⅓ cup of nuts count as 1 ounce of meat.

▲▲▲▲▲▲▲ 22 ▲▲▲▲▲▲▲

Glossary

beans   seeds that you can eat

muscles   body tissue connected to the bones that allows us to move

nuts   a hard-shelled seed that usually can be eaten

protein   a substance found in all living plant and animal cells

quail   a small, short-tailed bird with brown or gray feathers

vegetables   the edible part of a plant grown for food

About SandCastle™

A professional team of educators, reading specialists, and content developers created the SandCastle™ series to support young readers as they develop reading skills and strategies and increase their general knowledge. The SandCastle™ series has four levels that correspond to early literacy development in young children. The levels are provided to help teachers and parents select the appropriate books for young readers.

Emerging Readers
(no flags)

Beginning Readers
(1 flag)

Transitional Readers
(2 flags)

Fluent Readers
(3 flags)

These levels are meant only as a guide. All levels are subject to change.

To see a complete list of SandCastle™ books and other nonfiction titles from ABDO Publishing Company, visit **www.abdopub.com** or contact us at:

4940 Viking Drive, Edina, Minnesota 55435 • 1-800-800-1312 • fax: 1-952-831-1632